HOW I DISPROVED EINSTEIN TWICE

How I disproved Einstein Twice

Billy Coskun

How I Disproved Einstein Twice

Copyright © 2016 Billy Coskun

All rights reserved.

ISBN-10: 1543237290
ISBN-13: 978-1543237290

Cover designed by Billy Coskun

2 syncronized clocks

 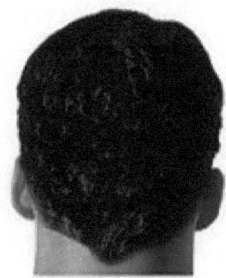 

L   **R**

In my reference frame clock L ticks faster than clock R

This image is being used in the Wikipedia article Diplacusis with permission given by the author Billy Coskun.

DEDICATION

I dedicate this book to my parents.

Nature is a work created by a conscious artist, and the artist is God.

- BILLY COSKUN

CONTENTS

INTRODUCTION

1   Chapter One   1

    Universe

    Big Bang

    Time concept

    The concept of now

    Space-time

2   Chapter Two   13

    What is light?

    Speed of light

    Gravity

3   Chapter Three   21

    Relativity briefly explained

    Relativity of simultaneity

    Time Dilation

    Gravitational Time Dilation

    Is Time Travel Possible?

    Einstein's Special Relativity and Spacetime theories are flawed

    CONCLUSION   43

    ATTRIBUTION   47

INTRODUCTION

This book is a shorter version of my book titled "The Holographic Universe", which is available on Amazon as an eBook and in paperback.

I have devised this book into three chapters. In the first two chapters, I will give information about the main physics subjects related to the contex of this book and the basis of my arguments. The third chapter is where things get really serious as I both logically and mathematically disprove Einstein's Special and General Relativity that the Space-time theory is based on. You are welcome to search physics textbooks to find the original theories and their mathematical expressions. I am confident that if you examine them carefully you will see the brilliancy in my reasoning. I think Einstein was very busy when he was developing his theories and put them out there without carefully reasoning. The scientific community took them seriously and nobody had the courage to question the validity of Einstein's theories, that is until now! Enjoy reading.

Billy Coskun

1 CHAPTER ONE

Universe

The Universe can be defined as everything that exists, everything that has existed, and everything that will exist. According to our current understanding, the Universe consists of space-time, forms of energy (including electromagnetic radiation and matter), and the physical laws that relate them. The Universe encompasses all of life, all of history, and some philosophers and scientists suggest that it even encompasses ideas such as mathematics and logic.

The Universe is all of time and space and its contents. The Universe includes planets, natural satellites, minor planets, stars, galaxies, the contents of intergalactic space, the smallest subatomic particles, and all matter and energy. The observable universe is about 28 billion parsecs (91 billion light-years) in diameter at the present time. The size of the whole Universe is not known and may be either finite or infinite. Observations and the development of physical theories have led to inferences about the composition and evolution of the Universe.

Throughout recorded history, cosmologies and cosmogonies, including scientific models, have been proposed to explain observations of the Universe. The earliest quantitative geocentric models were developed by ancient Greek philosophers and Indian philosophers. Over the centuries, more precise astronomical observations led to the Solar System and, Johannes Kepler's improvement on that model with elliptical orbits, which was eventually explained by Isaac Newton's theory of gravity. Further observational improvements led to the realization that the Solar System is located in a galaxy composed of billions of stars, the Milky Way. It was subsequently discovered that our galaxy is just one of many. On the largest scales, it is assumed that the distribution of galaxies is uniform and the same in all directions, meaning that the Universe has neither an edge nor a center. Observations of the distribution of these galaxies and their spectral lines have led to many of the theories of modern physical cosmology. The discovery in the early 20th century that galaxies are systematically redshifted suggested that the Universe is expanding, and the discovery of the cosmic microwave background radiation suggested that the Universe had a beginning. Finally, observations in the late 1990s indicated the rate of the expansion of the Universe is increasing indicating that the majority of energy is most likely in an unknown form called dark energy. The majority of mass in the universe also appears to exist in an unknown form, called dark matter.

There are many competing hypotheses about the ultimate fate of the Universe. Physicists and philosophers remain unsure about what, if anything, preceded the Big Bang. Many refuse to speculate, doubting that any information from any such prior state could ever be accessible. There are various multiverse hypotheses, in which some physicists have suggested that the Universe might be one among many universes that likewise exist.

Big Bang

The Big Bang theory is a cosmological model that is describing the development of the Universe. Space and time were created in the Big Bang, and these were merged with a fixed amount of energy and matter; as space expands, the density of that matter and energy decreases. After the initial expansion, the Universe cooled sufficiently to allow the formation first of subatomic particles and later of simple atoms. Giant clouds of these primordial elements later combined through gravity to form stars and galaxies. Some estimates place this moment at approximately 13.8 billion years ago, which is thus considered the age of the universe.

The Big Bang theory accounts for the fact that the universe expanded from a very high density and high temperature state, and offers a comprehensive explanation for a broad range of phenomena, including the abundance of light elements, the cosmic microwave background and large scale structure. If the known laws of physics are extrapolated beyond where they have been verified, there is a singularity.

Since Belgian priest, astronomer and professor of physics Georges Lemaître first noted in 1927, that an expanding universe might be traced back in time to an originating single point, scientists have built on his idea of cosmic expansion. While the scientific community was once divided between supporters of two different expanding universe theories, the Big Bang and the Steady State theory, accumulated empirical evidence provides strong support for the former. In 1929, from analysis of galactic redshifts, Edwin Hubble concluded that galaxies are drifting apart; this is important observational evidence consistent with the hypothesis of an expanding universe. In 1965, the cosmic microwave background radiation was discovered, which was crucial evidence in favor of the Big Bang model, since that theory predicted the existence of background radiation throughout the universe before it was discovered. More recently, measurements of the redshifts of supernovae indicate that the expansion of the universe is accelerating, an observation attributed to dark energy's existence. The known physical laws of nature can be used to calculate the characteristics of the universe in detail back in time to an initial state of extreme density and temperature.

American astronomer Edwin Hubble observed that the distances to faraway galaxies were strongly correlated with their redshifts. This was interpreted to mean that all distant galaxies and clusters are receding away from our vantage point with an apparent velocity proportional to their distance: that is, the farther they are, the faster they move away from us, regardless of direction. Assuming the Copernican principle (that the Earth is not the center of the universe), the only remaining interpretation is that all observable regions of the universe are receding from all others. Since we know that the distance between galaxies increases today, it must mean that in the past galaxies were closer together. The continuous expansion of the universe implies that the universe was denser and hotter in the past.

Large particle accelerators can replicate the conditions that prevailed after the early moments of the universe, resulting in confirmation and refinement of the details of the Big Bang model. However, these accelerators can only probe so far into high energy regimes. Consequently, the state of the universe in the earliest instants of the Big Bang expansion is still poorly understood and an area of open investigation and speculation.

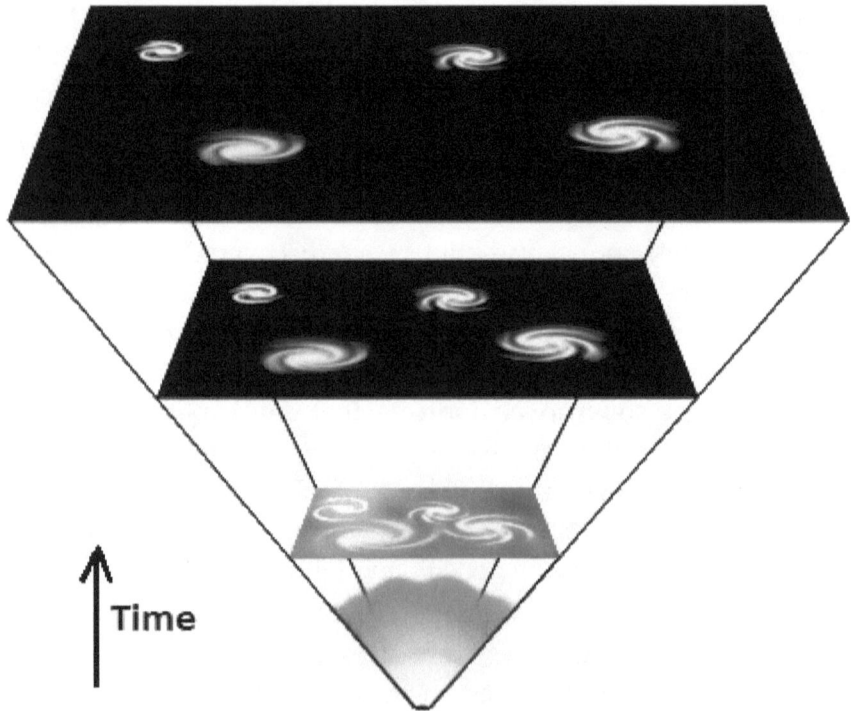

Figure 1.1 According to the Big Bang model, the universe expanded from an extremely dense and hot state and continues to expand.

The first subatomic particles include protons, neutrons, and electrons. Though simple atomic nuclei formed within the first three minutes after the Big Bang, thousands of years passed before the first electrically neutral atoms formed. The majority of atoms produced by the Big Bang were hydrogen, along with helium and traces of lithium. Giant clouds of these primordial elements later coalesced through gravity to form stars and galaxies, and the heavier elements were synthesized either within stars or during supernovae.

The framework for the Big Bang model relies on Albert Einstein's theory of general relativity and on simplifying assumptions such as homogeneity and isotropy of space. The governing equations were formulated by Alexander Friedmann, and similar solutions were worked on by Willem de Sitter. Since then, astrophysicists have incorporated observational and theoretical additions into the Big Bang model, and its parametrization as the Lambda-CDM model serves as the framework for current investigations of theoretical cosmology.

The Lambda-CDM model is the standard model of Big Bang cosmology, the simplest model that provides a reasonably good account of various observations about the universe with following properties of the cosmos:

The existence and structure of the cosmic microwave background

The large-scale structure in the distribution of galaxies

The abundances of hydrogen (including deuterium), helium, and lithium

The accelerating expansion of the universe observed in the light from distant galaxies and supernovae

The model assumes that general relativity is the correct theory of gravity on cosmological scales. It emerged in the late 1990s as a concordance cosmology, after a period of time when disparate observed properties of the universe appeared mutually inconsistent, and there was no consensus on the makeup of the energy density of the universe.

The Lambda-CDM model can be extended by adding cosmological inflation, quintessence and other elements that are current areas of speculation and research in cosmology.

Some alternative models challenge the assumptions of the Lambda-CDM model. Examples of these are modified Newtonian dynamics, modified gravity and theories of large-scale variations in the matter density of the universe.

Time Concept

Time is the indefinite continued progression of existence and events that occur in apparently irreversible succession from the past through the present to the future. Time is a component quantity of various measurements used to sequence events, to compare the duration of events or the intervals between them, and to quantify rates of change of quantities in material reality or in the conscious experience. Time is often referred to as the fourth dimension, along with the three spatial dimensions.

Time has long been a major subject of study in religion, philosophy, and science, but defining it in a manner applicable to all fields without circularity has consistently eluded scholars. Two contrasting viewpoints on time divide many prominent philosophers. One view is that time is part of the fundamental structure of the universe—a dimension independent of events, in which events occur in sequence. Isaac Newton subscribed to this realist view, and hence it is sometimes referred to as Newtonian time. The opposing view is that time does not refer to any kind of container that events and objects move through, nor to any entity that flows, but that it is instead part of a fundamental intellectual structure (together with space and number) within which humans sequence and compare events. This second view, in the tradition of Gottfried Leibniz and Immanuel Kant, holds that time is neither an event nor a thing, and thus is not itself measurable nor can it be travelled.

Time is one of the seven fundamental physical quantities in both the International System of Units and International System of Quantities. Time is used to define other quantities—such as velocity—so defining time in terms of such quantities would result in circularity of definition. An operational definition of time, wherein one says that observing a certain number of repetitions of one or another standard cyclical event (such as the passage of a free-swinging pendulum) constitutes one standard unit such as the second, is highly useful in the conduct of both advanced experiments and everyday affairs of life. The operational definition leaves aside the question whether there is something called time, apart from the counting activity just mentioned, that flows and that can be measured. Investigations of a single continuum called spacetime bring questions about space into questions about time, questions that have their roots in the works of early students of natural philosophy.

Furthermore, it may be that there is a subjective component to time, but whether or not time itself is felt, as a sensation, or is a judgment, is a matter of debate.

Figure 1.2 The flow of sand in an hourglass can be used to keep track of elapsed time. It also concretely represents the present as being between the past and the future.

Temporal measurement has occupied scientists and technologists, and was a prime motivation in navigation and astronomy. Periodic events and periodic motion have long served as standards for units of time. Examples include the apparent motion of the sun across the sky, the phases of the moon, the swing of a pendulum, and the beat of a heart. Currently, the international unit of time, the second, is defined by measuring the electronic transition frequency of caesium. Time is also of significant social importance, having economic value as well as personal value, due to an awareness of the limited time in each day and in human life spans.

The concept of now

The present (or here and now) is the time that is associated with the events perceived directly and in the first time, not as a recollection (perceived more than once) or a speculation. It is a period of time between the past and the future, and can vary in meaning from being an instant to a day or longer.

It is sometimes represented as a hyperplane in space-time, typically called "now", although modern physics demonstrates that such a hyperplane cannot be defined uniquely for observers in relative motion. The present may also be viewed as a duration.

Albert Einstein's Special Theory of Relativity postulates that there is no such thing as absolute simultaneity. When care is taken to operationalise the present, it follows that the events that can be labeled as simultaneous with a given event, can not be in direct cause-effect relationship. Such collections of events are perceived differently by different observers. Instead, when focusing on "now" as the events perceived directly, not as a recollection or a speculation, for a given observer "now" takes the form of the observer's past light cone. The light cone of a given event is objectively defined as the collection of events in causal relationship to that event, but each event has a different associated light cone. One has to conclude that in relativistic models of physics there is no place for "the present" as an absolute element of reality. Einstein phrased this as: "People like us, who believe in physics, know that the distinction between past, present, and future is only a stubbornly persistent illusion".

In physical cosmology, the present time in the chronology of the universe is estimated at 13.8 billion years after the singularity determining the arrow of time. In terms of the metric expansion of space, it is in the dark-energy-dominated era, after the universe's matter content has become diluted enough for metric expansion to be dominated by vacuum energy (dark energy). It is also in the universe's Stelliferous Era, after enough time for superclusters to have formed (at about 5 billion years), but before the accelerating expansion of the universe has removed the local supercluster beyond the cosmological horizon (at about 150 billion years).

Spacetime

In physics, spacetime is any mathematical model that combines space and time into a single interwoven continuum. The spacetime of our universe has historically been interpreted from a Euclidean space perspective, which regards space as consisting of three dimensions, and time as consisting of one dimension, the fourth dimension. By combining space and time into a single manifold called Minkowski space, physicists have significantly simplified a large number of physical theories, as well as described in a more uniform way the workings of the universe at both the supergalactic and subatomic levels.

The basic elements of spacetime are events. In any given spacetime, an event is a unique position at a unique time. Because events are spacetime points, an example of an event in classical relativistic physics is the location of an elementary particle at a particular time. A spacetime itself can be viewed as the union of all events in the same way that a line is the union of all of its points, formally organized into a manifold, a space which can be described at small scales using coordinate systems.

A spacetime is independent of any observer. However, in describing physical phenomena which occur at certain moments of time in a given region of space, each observer chooses a convenient metrical coordinate system. Events are specified by four real numbers in any such coordinate system. The trajectories of elementary (point-like) particles through space and time are thus a continuum of events called the world line of the particle. Extended or composite objects consisting of many elementary particles are thus a union of many world lines twisted together by virtue of their interactions through spacetime into a world-braid.

However, in physics, it is common to treat an extended object as a particle or field with its own unique (e.g., center of mass) position at any given time, so that the world line of a particle or light beam is the path that this particle or beam takes in the spacetime and represents the history of the particle or beam. The world line of the orbit of the Earth in such a description is depicted in two spatial dimensions x and y (the plane of the Earth's orbit) and a time dimension orthogonal to x and y. The orbit of the Earth is an ellipse in space alone, but its world line is a helix in spacetime.

In non-relativistic classical mechanics, the use of Euclidean space instead of spacetime is appropriate, because time is treated as universal with a constant rate of passage that is independent of the state of motion of an observer. In relativistic contexts, time cannot be separated from the three dimensions of space, because the observed rate at which time passes for an object depends on the object's velocity relative to the observer and also on the strength of gravitational fields, which can slow the passage of time for an object as seen by an observer outside the field.

Until the beginning of the 20th century, time was believed to be independent of motion, progressing at a fixed rate in all reference frames; however, following its prediction by special relativity, later experiments confirmed that time slows at higher speeds of the reference frame relative to another reference frame. Such slowing, called time dilation, is explained in special relativity theory. Many experiments have confirmed time dilation, such as the relativistic decay of muons from cosmic ray showers and the slowing of atomic clocks aboard a Space Shuttle relative to synchronized Earth-bound inertial clocks. The duration of time can therefore vary according to events and reference frames.

The term spacetime has taken on a generalized meaning beyond treating spacetime events with the normal 3+1 dimensions. It is really the combination of space and time. Other proposed spacetime theories include additional dimensions - normally spatial but there exist some speculative theories that include additional temporal dimensions and even some that include dimensions that are neither temporal nor spatial (e.g., superspace). How many dimensions are needed to describe the universe is still an open question. Speculative theories such as string theory predict 10 or 26 dimensions (with M-theory predicting 11 dimensions: 10 spatial and 1 temporal), but the existence of more than four dimensions would only appear to make a difference at the subatomic level.

2 CHAPTER TWO

What is light?
Light is electromagnetic radiation within a certain portion of the electromagnetic spectrum. The word usually refers to visible light, which is visible to the human eye and is responsible for the sense of sight.

The main source of light on Earth is the Sun. Sunlight provides the energy that green plants use to create sugars mostly in the form of starches, which release energy into the living things that digest them. This process of photosynthesis provides virtually all the energy used by living things. Another important source of light for humans has been fire, from ancient campfires to modern kerosene lamps. With the development of electric lights and power systems, electric lighting has effectively replaced firelight. Some species of animals generate their own light, a process called bioluminescence. For example, fireflies use light to locate mates, and vampire squids use it to hide themselves from prey.

In physics, the term light sometimes refers to electromagnetic radiation of any wavelength, whether visible or not. In this sense, gamma rays, X-rays, microwaves and radio waves are also light. Like all types of light, visible light is emitted and absorbed in tiny packets called photons and exhibits properties of both waves and particles. This property is referred to as the wave–particle duality. The study of light, known as optics, is an important research area in modern physics.

Figure 2.1 A cloud illuminated by sunlight

Speed of light

The speed of light, commonly denoted c, is a universal physical constant important in many areas of physics. Its value is approximately 300 km/s, since the length of the meter is defined from this constant and the international standard for time. According to special relativity, c is the maximum speed at which all matter and information in the universe can travel. It is the speed at which all massless particles and changes of the associated fields (including electromagnetic radiation such as light and gravitational waves) propagate. Such particles and waves travel at c regardless of the motion of the source or the inertial reference frame of the observer. In the theory of relativity, c interrelates space and time, and also appears in the famous Einstein equation of mass–energy equivalence $E = mc2$.

For many practical purposes, light and other electromagnetic waves will appear to propagate instantaneously, but for long distances and very sensitive measurements, their finite speed has noticeable effects. In communicating with distant space probes, it can take minutes to hours for a message to get from Earth to the spacecraft, or vice versa. The light seen from stars left them many years ago, allowing the study of the history of the universe by looking at distant objects (Alpha Centauri is the closest star system to the Solar System at a distance of 4.37 light-years). The finite speed of light also limits the theoretical maximum speed of computers, since information must be sent within the computer from chip to chip. The speed of light can be used with time of flight measurements to measure large distances to high precision.

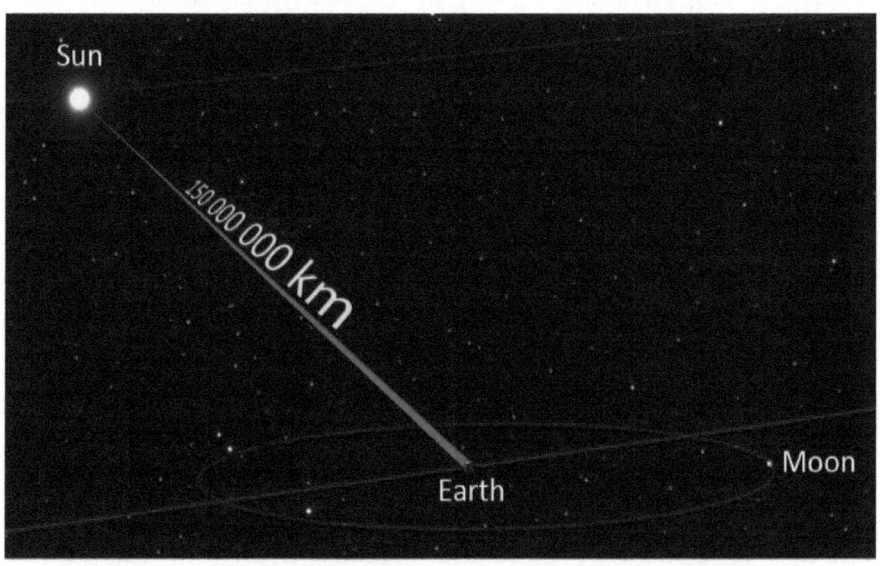

Figure 2.2 Sunlight takes about 8 minutes 17 seconds to travel the average distance from the surface of the Sun to the Earth.

Danish astronomer Ole Rømer first demonstrated in 1676 that light travels at a finite speed (as opposed to instantaneously) by studying the apparent motion of Jupiter's moon Io. In 1865, Scottish scientist James Clerk Maxwell proposed that light was an electromagnetic wave, and therefore travelled at the speed c appearing in his theory of electromagnetism. In 1905, Albert Einstein postulated that the speed of light c with respect to any inertial frame is a constant and is independent of the motion of the light source. He explored the consequences of that postulate by deriving the special theory of relativity and in doing so showed that the parameter c had relevance outside of the context of light and electromagnetism.

Gravity

Gravity or gravitation is a natural phenomenon by which all things with energy are brought toward one another, including stars, planets, galaxies and even light and sub-atomic particles. Gravity is responsible for many of the structures in the Universe, by creating spheres of hydrogen — where hydrogen fuses under pressure to form stars — and grouping them into galaxies. On Earth, gravity gives weight to physical objects and causes the tides. Gravity has an infinite range, although its effects become increasingly weaker on farther objects.

Gravity is most accurately described by the general theory of relativity (proposed by Albert Einstein in 1915) which describes gravity not as a force but as a consequence of the curvature of spacetime caused by the uneven distribution of mass/energy; and resulting in gravitational time dilation, where time lapses more slowly in stronger gravitational potential. However, for most applications, gravity is well approximated by Newton's law of universal gravitation, which postulates that gravity causes a force where two bodies of mass are directly drawn to each other according to a mathematical relationship, where the attractive force is proportional to the product of their masses and inversely proportional to the square of the distance between them. This is considered to occur over an infinite range, such that all bodies with mass in the universe are drawn to each other no matter how far they are apart.

Gravity is the weakest of the four fundamental interactions of nature (the other three are the strong force, the electromagnetic force, and the weak force). As a consequence, gravity has a negligible influence on the behavior of sub-atomic particles, and plays no role in determining the internal properties of everyday matter. On the other hand, gravity is the dominant interaction at the macroscopic scale, and is the cause of the formation, shape, and trajectory (orbit) of astronomical bodies. It is responsible for various phenomena observed on Earth and throughout the universe; for example, it causes the Earth and the other planets to orbit the Sun, the Moon to orbit the Earth, the formation of tides, and the formation and evolution of galaxies, stars and the Solar System.

In pursuit of a theory of everything, the merging of general relativity and quantum mechanics (or quantum field theory) into a more general theory of quantum gravity has become an area of research.

Modern work on gravitational theory began with the work of Galileo Galilei in the late 16th and early 17th centuries. In his famous (though possibly apocryphal) experiment dropping balls from the Tower of Pisa, and later with careful measurements of balls rolling down inclines, Galileo showed that gravity accelerates all objects at the same rate. This was a major departure from Aristotle's belief that heavier objects accelerate faster. Galileo postulated air resistance as the reason that lighter objects may fall more slowly in an atmosphere. Galileo's work set the stage for the formulation of Newton's theory of gravity.

Newton's theory enjoyed its greatest success when it was used to predict the existence of Neptune based on motions of Uranus that could not be accounted for by the actions of the other planets. Calculations by both John Couch Adams and Urbain Le Verrier predicted the general position of the planet, and Le Verrier's calculations are what led Johann Gottfried Galle to the discovery of Neptune.

A discrepancy in Mercury's orbit pointed out flaws in Newton's theory. By the end of the 19th century, it was known that its orbit showed slight perturbations that could not be accounted for entirely under Newton's theory, but all searches for another perturbing body (such as a planet orbiting the Sun even closer than Mercury) had been fruitless. The issue was resolved in 1915 by Albert Einstein's new theory of general relativity, which accounted for the small discrepancy in Mercury's orbit.

Although Newton's theory has been superseded by the Einstein's general relativity, most modern non-relativistic gravitational calculations are still made using the Newton's theory because it is simpler to work with and it gives sufficiently accurate results for most applications involving sufficiently small masses, speeds and energies.

In general relativity, the effects of gravitation are ascribed to spacetime curvature instead of a force. The starting point for general relativity is the equivalence principle, which equates free fall with inertial motion and describes free-falling inertial objects as being accelerated relative to non-inertial observers on the ground. In Newtonian physics, however, no such acceleration can occur unless at least one of the objects is being operated on by a force.

Einstein proposed that spacetime is curved by matter, and that free-falling objects are moving along locally straight paths in curved spacetime. These straight paths are called geodesics. Like Newton's first law of motion, Einstein's theory states that if a force is applied on an object, it would deviate from a geodesic. For instance, we are no longer following geodesics while standing because the mechanical resistance of the Earth exerts an upward force on us, and we are non-inertial on the ground as a result. This explains why moving along the geodesics in spacetime is considered inertial.

English astronomer, physicist, and mathematician Arthur Eddington conducted an expedition to observe the Solar eclipse of 29 May 1919 that provided one of the earliest confirmations of General Relativity, and he became known for his popular expositions and interpretations of the theory. During World War I, Eddington was Secretary of the Royal Astronomical Society, which meant he was the first to receive a series of letters and papers from Willem de Sitter regarding Einstein's theory of general relativity. Eddington was fortunate in being not only one of the few astronomers with the mathematical skills to understand general relativity, but owing to his internationalist and pacifist views inspired by his Quaker religious beliefs, one of the few at the time who was still interested in pursuing a theory developed by a German physicist. He quickly became the chief supporter and expositor of relativity in Britain. He and Astronomer Royal Frank Watson Dyson organized two expeditions to observe a solar eclipse in 1919 to make the first empirical test of Einstein's theory: the measurement of the deflection of light by the sun's gravitational field.

Figure 2.3 One of Eddington's photographs of the total solar eclipse of 29 May 1919, presented in his 1920 paper announcing its success, confirming Einstein's theory that light "bends"

Eddington travelled to the island of Príncipe near Africa to watch the solar eclipse of 29 May 1919. During the eclipse, he took pictures of the stars (several stars in the Hyades cluster include Kappa Tauri of the constellation Taurus) in the region around the Sun. According to the theory of general relativity, stars with light rays that passed near the Sun would appear to have been slightly shifted because their light had been curved by its gravitational field. This effect is noticeable only during eclipses, since otherwise the Sun's brightness obscures the affected stars. Eddington showed that Newtonian gravitation could be interpreted to predict half the shift predicted by Einstein.

Eddington's observations published the next year confirmed Einstein's theory, and were hailed at the time as a conclusive proof of general relativity over the Newtonian model. The news was reported in newspapers all over the world as a major story. Afterward, Eddington embarked on a campaign to popularize relativity and the expedition as landmarks both in scientific development and international scientific relations.

It has been claimed that Eddington's observations were of poor quality, and he had unjustly discounted simultaneous observations at Sobral, Brazil, which appeared closer to the Newtonian model, but a 1979 re-analysis with modern measuring equipment and contemporary software validated Eddington's results and conclusions. The quality of the 1919 results was indeed poor compared to later observations, but was sufficient to persuade contemporary astronomers. The rejection of the results from the Brazil expedition was due to a defect in the telescopes used which, again, was completely accepted and well-understood by contemporary astronomers.

In the decades after the discovery of general relativity, it was realized that general relativity is incompatible with quantum mechanics. It is possible to describe gravity in the framework of quantum field theory like the other fundamental forces, such that the attractive force of gravity arises due to exchange of virtual gravitons, in the same way as the electromagnetic force arises from exchange of virtual photons. This reproduces general relativity in the classical limit. However, a more complete theory of quantum gravity (or a new approach to quantum mechanics) is required.

3 CHAPTER THREE

Relativity Briefly Explained

The theory of relativity, or simply relativity in physics, usually encompasses two theories by Albert Einstein: special relativity and general relativity. Concepts introduced by the theories of relativity include spacetime as a unified entity of space and time, relativity of simultaneity, kinematic and gravitational time dilation, and length contraction. The theory of relativity transformed theoretical physics and astronomy during the 20th century. When first published, relativity superseded a 200-year-old theory of mechanics created primarily by Isaac Newton.

Special relativity is a theory of the structure of spacetime. It was introduced in Einstein's 1905 paper "On the Electrodynamics of Moving Bodies". Special relativity is based on two postulates which are contradictory in classical mechanics:

The laws of physics are the same for all observers in uniform motion relative to one another (principle of relativity). The speed of light in a vacuum is the same for all observers, regardless of their relative motion or of the motion of the light source.

The theory has many surprising and counter-intuitive consequences. Some of these are:

Relativity of simultaneity: Two events, simultaneous for one observer, may not be simultaneous for another observer if the observers are in relative motion.

Time dilation: Moving clocks are measured to tick more slowly than an observer's "stationary" clock.

Relativistic mass.

Length contraction: Objects are measured to be shortened in the direction that they are moving with respect to the observer.

Mass–energy equivalence: E = mc², energy and mass are equivalent and transmutable.

Maximum speed is finite: No physical object, message or field line can travel faster than the speed of light in a vacuum.

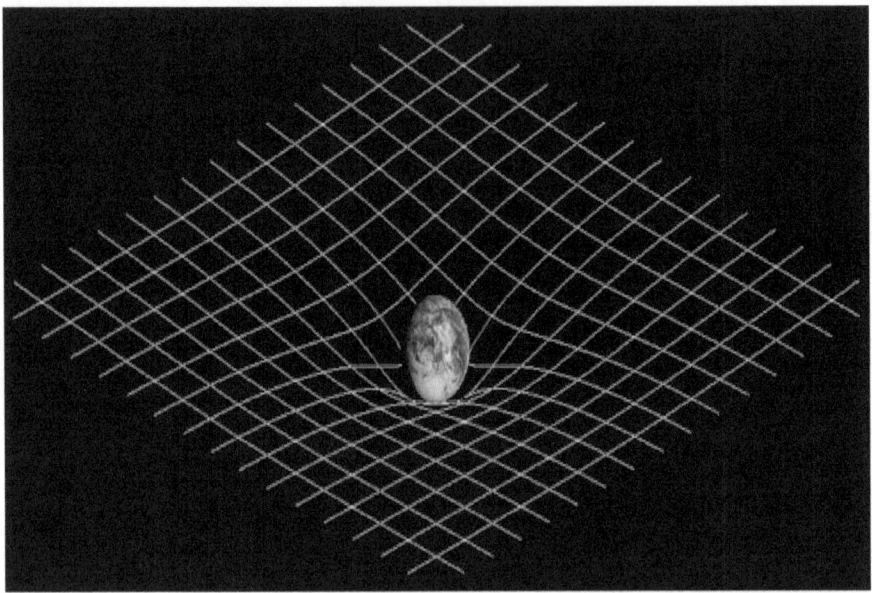

Figure 3.1 Two-dimensional projection of a three-dimensional analogy of spacetime curvature described in general relativity

General relativity has emerged as a highly successful model of gravitation and cosmology, which has so far passed many unambiguous observational and experimental tests. However, there are strong indications the theory is incomplete. The problem of quantum gravity and the question of the reality of spacetime singularities remain open. Observational data that is taken as evidence for dark energy and dark matter could indicate the need for new physics. Even taken as is, general relativity is rich with possibilities for further exploration. Mathematical relativists seek to understand the nature of singularities and the fundamental properties of Einstein's equations, and increasingly powerful computer simulations (such as those describing merging black holes) are run. In February 2016, it was announced that the existence of gravitational waves was directly detected by the Advanced LIGO team. A century after its publication, general relativity remains a highly active area of research. If general relativity were considered to be one of the two pillars of modern physics, then quantum theory, the basis of understanding matter from elementary particles to solid state physics, would be the other. However, how to reconcile quantum theory with general relativity is still an open question.

Relativity of Simultaneity

The relativity of simultaneity is the concept that distant simultaneity – whether two spatially separated events occur at the same time – is not absolute, but depends on the observer's reference frame. According to the special theory of relativity, it is impossible to say in an absolute sense that two distinct events occur at the same time if those events are separated in space. For example, a car crash in London and another in New York, which appear to happen at the same time to an observer on the earth, will appear to have occurred at slightly different times to an observer on an airplane flying between London and New York. The question of whether the events are simultaneous is relative: in the stationary earth reference frame the two accidents may happen at the same time but in other frames (in a different state of motion relative to the events) the crash in London may occur first, and in still other frames the New York crash may occur first. However, if the two events could be causally connected (i.e. the time between event A and event B is greater than the distance between them divided by the speed of light), the order is preserved (i.e., "event A precedes event B") in all frames of reference. If we imagine one reference frame assigns precisely the same time to two events that are at different points in space, a reference frame that is moving relative to the first will generally assign different times to the two events.

A popular picture for understanding this idea is provided by a thought experiment consisting of one observer midway inside a speeding traincar and another observer standing on a platform as the train moves past. It is similar to thought experiments suggested by Daniel Frost Comstock in 1910 and Einstein in 1917.

A flash of light is given off at the center of the traincar just as the two observers pass each other. The observer on board the train sees the front and back of the traincar at fixed distances from the source of light and as such, according to this observer, the light will reach the front and back of the traincar at the same time.

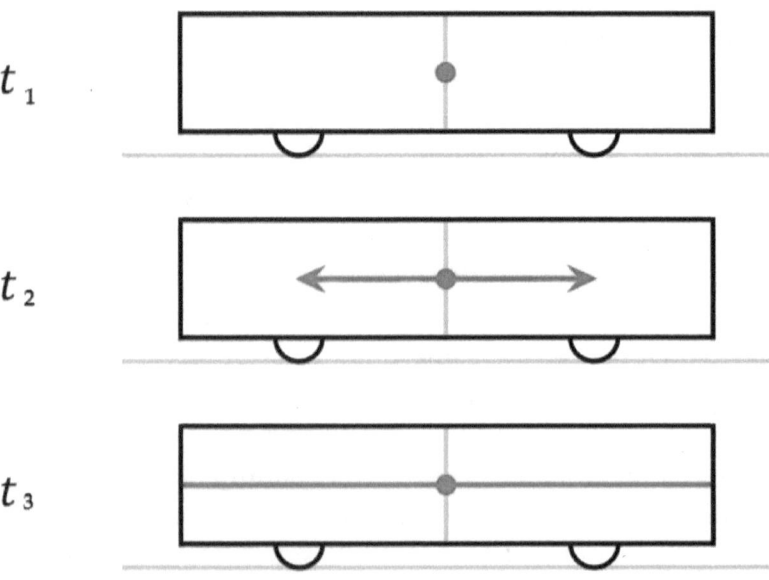

Figure 3.2 The train-and-platform experiment from the reference frame of an observer on board the train (*t* represents time)

The observer standing on the platform, on the other hand, sees the rear of the traincar moving toward the point at which the flash was given off and the front of the traincar moving away from it. As the speed of light is finite and the same in all directions for all observers, the light headed for the back of the train will have less distance to cover than the light headed for the front. Thus, the flashes of light will strike the ends of the traincar at different times.

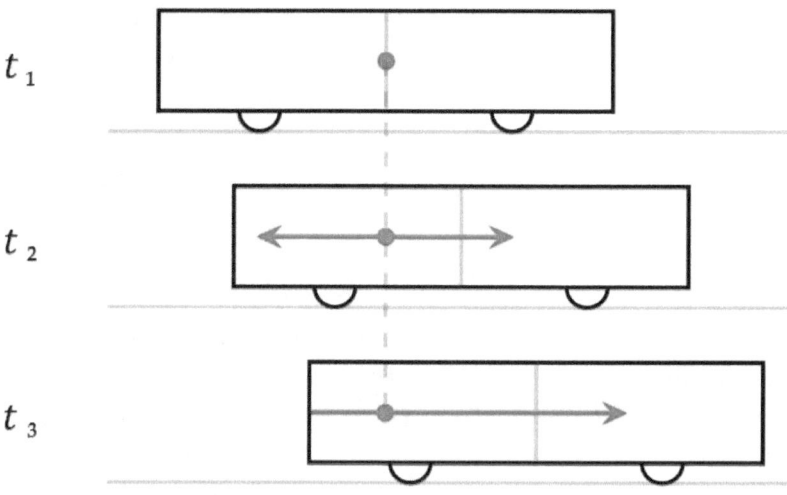

Figure 3.3 Reference frame of an observer standing on the platform

Einstein's version of the experiment presumed slightly different conditions, where a train moving past the standing observer is struck by two bolts of lightning simultaneously, but at different positions along the axis of train movement (back and front of the train car). In the inertial frame of the standing observer, there are three events which are spatially dislocated, but simultaneous: event of the standing observer facing the moving observer, event of lightning striking the front of the train car, and the event of lightning striking the back of the car.

Since the events are placed along the axis of train movement, their time coordinates become projected to different time coordinates in the moving train's inertial frame. Events which occurred at space coordinates in the direction of train movement (in the stationary frame), happen earlier than events at coordinates opposite to the direction of train movement. In the moving train's inertial frame, this means that lightning will strike the front of the train car before two observers align (face each other).

My argument: "The observer on board the train sees the front and back of the train car at fixed distances from the source of light and as such, according to this observer, the light will reach the front and back of the train car at the same time" statement postulated by the theory of relativity is not a valid statement. Because the light beams reflected from the front and the back of the train car will reach the moving observer in the middle of the train car at different times. The observer is moving toward the front of the car and moving away from the back when the emitted light beam is reflected. The light reflected from the front of the car will take less time to reach the observer than the light beam reflected from the back. That said, the observer will see the flashes of light located in the middle of the train car strike the back of the train car at the exact same time, because even though the light will reach the back of the car faster, it has to be reflected back for the observer to see it, so there will be an added time delay since the observer is moving away from the back of the train car.

My reasoning shows that Relativity of Simultaneity which is based on Einstein's Special Relativity is fundamentally an incomplete (therefore invalid) theory, because the flash of light given off at the center of the train car will reach the back of the train car before it reaches the front (same for both observers). This is very important, because if you have noticed both Comstock and Einstein failed to add the time delay caused by the light reflected from the back and the front to reach the observer standing in the middle of the train car. If you add this delay and make accurate calculations you will obtain the same result for both observers. This means that the relativity issue is not a physics phenomena. There is nothing new here. Newton's laws covered everything related to motion.

My own thought experiment: Let's say there are two light bulbs, each placed at the back and front of the train car. The bulbs are connected to their own portable power source and set to simultaneously flash at certain intervals. If the train is not moving, the light beams emitted by the bulbs will reach the observer standing in the middle of the car at the same time. If the train starts to move and reaches a constant speed of v, the beam emitted by the light bulb at the front of the car will reach the observer before the light beam emitted by the other bulb placed at the back, even though the observer **knows** that the light bulbs flash simultaneously. My thought experiment demonstrates that the Relativity issue is over-rated. Even an email sent from a computer takes a time to reach its recipient. Therefore, it is in my opinion that most aspects of the Theory of Relativity are just nonsense and as such, I consider the theory as one of the most unfortunate developments in physics science. That said, the maximum speed the train can move will be bound by the speed of light, because if the train moves towards the speed of light, the light beam from the front bulb will take less and less time to reach the observer. The speed of the train cannot exceed the speed of light because the event of the flashing of light hasn't happened yet. Einstein postulated that an infinite amount of energy would be required (which makes sense) for the train to move faster than the speed of light, therefore it's not possible.

Time Dilation

In the theory of relativity, time dilation is a difference of elapsed time between two events as measured by observers either moving relative to each other or differently situated from a gravitational mass or masses.

A clock at rest with respect to one observer may be measured to tick at a different rate when compared to a second observer's clock. This effect arises neither from technical aspects of the clocks nor from the propagation time of signals, but from the nature of spacetime.

Time dilation can be inferred from the observed constancy of the speed of light in all reference frames. This constancy of the speed of light means, counter to intuition, that speeds of material objects and light are not additive. It is not possible to make the speed of light appear greater by approaching at speed towards the material source that is emitting light. It is not possible to make the speed of light appear less by receding from the source at speed. From one point of view, it is the implications of this unexpected constancy that take away from constancies expected elsewhere. From the frame of reference of a moving observer traveling at the speed v relative to the rest frame of the clock, the light pulse traces out a longer, angled path.

The second postulate of special relativity states that the speed of light in free space is constant for all inertial observers, which implies a lengthening of the period of this clock from the moving observer's perspective. That is to say, in a frame moving relative to the clock, the clock appears to be running more slowly which expresses the fact that for the moving observer the period of the clock is longer than in the frame of the clock itself.

When two observers are in relative uniform motion and uninfluenced by any gravitational mass, the point of view of each will be that the other's moving clock is ticking at a slower rate than the local clock. The faster the relative velocity, the greater the magnitude of time dilation. This case is sometimes called special relativistic time dilation.

For instance, two rocket ships (A and B) speeding past one another in space would experience time dilation. If they somehow had a clear view into each other's ships, each crew would see the others' clocks and movement as going more slowly. That is, inside the frame of reference of Ship A, everything is moving normally, but everything over on Ship B appears to be moving more slowly (and vice versa).

From a local perspective, time registered by clocks that are at rest with respect to the local frame of reference (and far from any gravitational mass) always appears to pass at the same rate. In other words, if a new ship, Ship C, travels alongside Ship A, it is at rest relative to Ship A. From the point of view of Ship A, new Ship C's time would appear normal too.

In Albert Einstein's theory of relativity, time dilation in these two circumstances can be summarized:

In special relativity clocks,(hypothetically far from all gravitational mass) that are moving with respect to an inertial system of observation are measured to be running more slowly. This effect is described precisely by the Lorentz transformation.

In general relativity, clocks at a position with lower gravitational potential – such as in closer proximity to a planet – are found to be running more slowly. The articles on gravitational time dilation and gravitational redshift give a more detailed discussion.

Special and general relativistic effects can combine (as seen with ISS astronauts).

In special relativity, the time dilation effect is reciprocal: as observed from the point of view of either of two clocks which are in motion with respect to each other, it will be the other clock that is time dilated. This presumes that the relative motion of both parties is uniform; that is, they do not accelerate with respect to one another during the course of the observations. In contrast, gravitational time dilation (as treated in general relativity) is not reciprocal: an observer at the top of a tower will observe that clocks at ground level tick slower, and observers on the ground will agree about the direction and the magnitude of the difference. There is still some disagreement in a sense, because all the observers believe their own local clocks are correct, but the direction and ratio of gravitational time dilation is agreed by all observers, independent of their altitude.

Science fiction enthusiasts have noted the implications time dilation has on forward time travel, technically making it possible. The Hafele and Keating experiment involved flying planes around the world with atomic clocks on board. Upon the trips' completion the clocks were compared to a static, ground based atomic clock. It was found that 273±7 nanoseconds had been gained on the planes' clocks. The current human time travel record holder is Russian cosmonaut Sergei Krikalev, who beat the previous record of about 20 milliseconds by cosmonaut Sergei Avdeyev.

Time dilation would make it possible for passengers in a fast-moving vehicle to travel further into the future while aging very little, in that their great speed slows down the passage of on-board time relative to that of an observer. That is, the ship's clock (and according to relativity, any human traveling with it) shows less elapsed time than the clocks of observers on earth. For sufficiently high speeds the effect is dramatic. For example, one year of travel might correspond to ten years at home. Indeed, a constant 1 g acceleration would permit humans to travel through the entire known Universe in one human lifetime. The space travelers could return to Earth billions of years in the future. A scenario based on this idea was presented in the novel Planet of the Apes by Pierre Boulle.

A more likely use of this effect would be to enable humans to travel to nearby stars without spending their entire lives aboard a ship. However, any such application of time dilation during interstellar travel would require the use of some new, advanced method of propulsion. The Orion Project has been the only major attempt toward this idea.

Current space flight technology has fundamental theoretical limits based on the practical problem that an increasing amount of energy is required for propulsion as a craft approaches the speed of light. The likelihood of collision with small space debris and other particulate material is another practical limitation. At the velocities presently attained, however, time dilation occurs but is too small to be a factor in space travel. Travel to regions of spacetime where gravitational time dilation is taking place, such as within the gravitational field of a black hole but outside the event horizon (perhaps on a hyperbolic trajectory exiting the field), could also yield results consistent with present theory.

Gravitational time dilation

Gravitational time dilation is a form of time dilation, an actual difference of elapsed time between two events as measured by observers situated at varying distances from a gravitating mass. The weaker the gravitational potential (the farther the clock is from the source of gravitation), the faster time passes. Albert Einstein originally predicted this effect in his theory of relativity and it has since been confirmed by tests of general relativity.

This has been demonstrated by noting that atomic clocks at differing altitudes (and thus different gravitational potential) will eventually show different times. The effects detected in such Earth-bound experiments are extremely small, with differences being measured in nanoseconds. Demonstrating greater effects would require greater distances from the Earth or a larger gravitational source.

Gravitational time dilation was first described by Albert Einstein in 1907 as a consequence of special relativity in accelerated frames of reference. In general relativity, it is considered to be a difference in the passage of proper time at different positions as described by a metric tensor of spacetime. The existence of gravitational time dilation was first confirmed directly by the Pound–Rebka experiment in 1959.

Clocks that are far from massive bodies (or at lower gravitational potentials) run more quickly, and clocks close to massive bodies (or at higher gravitational potentials) run more slowly. For example, considered over the total lifetime of the earth (4.6 Gyr), a clock set at the peak of Mount Everest would be about 39 hours ahead of a clock set at sea level. This is because gravitational time dilation is manifested in accelerated frames of reference or, by virtue of the equivalence principle, in the gravitational field of massive objects.

According to general relativity, inertial mass and gravitational mass are the same, and all accelerated reference frames (such as a uniformly rotating reference frame with its proper time dilation) are physically equivalent to a gravitational field of the same strength.

Is Time Travel Possible?

Time travel is the concept of movement (such as by a human) between different points in space, typically using a hypothetical device known as a time machine, in the form of a vehicle or of a portal connecting distant points in time. Some theories, most notably special and general relativity, suggest that suitable geometries of spacetime or specific types of motion in space might allow time travel into the past and future if these geometries or motions were possible.

Figure 3.4 This atomic clock - flown westward around the world in a Boeing by Hafele & Keating in 1977 is permanently in the future by 273 nanoseconds. To me it's the proof that Spacetime theory is flawed.

Relativity predicts that if one were to move away from the Earth at relativistic velocities and return, more time would have passed on Earth than for the traveler, so in this sense it is accepted that relativity allows travel into the future (according to relativity there is no single objective answer to how much time has really passed between the departure and the return, but there is an objective answer to how much proper time has been experienced by both the Earth and the traveler, i.e., how much each has aged). With current technologies, it is only possible to cause a human traveler to age less than companions on Earth by a very small fraction of a second, the current record being about 20 milliseconds for the cosmonaut Sergei Avdeyev. However, as I explained in a theory, the Universal Quantum Clock events are same and simultaneous for the observer on earth and the astronauts in the space ship. In this sense, astronauts don't really travel to the future. They remain in the same quantum clock simultaneity with the observers on earth at all times. This is a self-proven proposition. Otherwise, astronauts would never be able to return to earth from the past (traveling into the future).

Many in the scientific community believe that backward time travel is highly unlikely. Any theory that would allow time travel would introduce potential problems of causality. The classic example of a problem involving causality is the grandfather paradox: what if one were to go back in time and kill one's own grandfather before one's father was conceived?

Of course, these are all hypothetical examples. Stephen Hawking has suggested that the absence of tourists from the future is an argument against the existence of time travel. This simply means that, until the time machine was actually to be invented, we would not be able to see time travelers.

On the other hand, according to Einstein's theory of general relativity, relative to the earth's age in billion of years, the earth's core is effectively 2.5 years younger than the surface. So, in other words, you can travel to the past if you can stand the heat!

Or, if you prefer traveling to the future and staying cool at the same time, you might want to climb to the peak of Mount Everest. Considered over the total lifetime of the earth (4.6 Gyr), a clock set at the peak of Mount Everest would be about 39 hours ahead of a clock set at sea level.

If Einstein's postulate to be considered seriously, we go through many different time frames during the course of our daily lives. This is caused by the change in the altitude and change in speed because of traveling from one point in space-time to another. For example, if you are video chatting with a friend on Skype while vacationing in Ibiza and your friend is in the Alps, you are seeing and talking to someone from the future.

It's very easy to disprove Einstein's general relativity. This is how: Let's say you are on the Mount Everest. You will be bored because there is nothing to do and you want to go home. This requires traveling back to the past. For this, you need to travel at near the speed of light. Or, you can call the nearest ground station for the climbers, they send a helicopter and return you back to the present time (!) safely - without the drama and danger of traveling near the speed of light which is 300.000 km/sec.

In short, Einstein's theory contradicts itself!

Another way to look at it would be by considering the paradox of earth's rotation. If the earth's core is younger, it means historically it rotated fewer times than the crust. Even if we dig a short well, we shouldn't be able to reach its bottom, because the bottom part hasn't happened yet in space-time! It's not even traveling to the future, the bottom part shouldn't exist, period!

Or, if there are 2 clocks placed on shelves on a wall vertically 1 meter apart from each other, they will eventually start to display different times because of difference in gravity. I shouldn't be able to reach and make adjustments to the clocks. Because, if I am at the same level with the clock on the bottom shelf, I am in the past time compared to the faster-ticking clock on the top shelf. The same thing is valid for the satellite clocks. We shouldn't be able to make constant corrections to the clock time on satellites by sending signals from the stations on earth for the purpose of maintaining the correct GPS data.

If you examine carefully, Einstein's theories are full of contradictions such as these.

Einstein's Special Relativity and Spacetime theories are flawed!

One of the important implications of special relativity of space-time theory is that for a moving person time passes more slowly. In other words, a traveler will age more slowly than a person who is not mobile. In special relativity, there is a thought experiment which is called the twin paradox involving identical twins, one of whom makes a journey into space in a high-speed rocket and returns home to find that the twin who remained on Earth has aged more. This is called paradox for a good reason because it makes no sense! I will elaborate my reasoning in more detail below but the postulation made by the theory of special relativity that "the time dilation effect is reciprocal: as observed from the point of view of either of two clocks which are in motion with respect to each other, it will be the other clock that is time dilated" is self-contradicting, to begin with. Even without deeper analyses, I think it's easy for anyone to realize the flaw in the theory when I conclude that it is practically impossible for the clocks to become both time-dilated at the same time with respect to each other. When the clocks brought to the same location, they will either show the same time or one clock will be ahead of the other. In the Hafele and Keating experiment, the atomic clock on the board of a plane that gained 273 nanoseconds compared to the static ground based atomic clock was the result of mechanical malfunctioning caused by the reduced gravity during the flight.

The idea of time slowing down for the traveler started with Einstein, who realized that further you move away from a stationary clock, it will take more time for the light reflected from the clock to reach you. I don't argue with this reasoning because it is basic geometry, but how reliable is the light as a time measuring medium really? Einstein postulated that, if a spaceship moves near the speed of light, the duration or time passed between the ticking of the clock which represents seconds for an astronaut on the spaceship can take years for another observer on earth! Spacetime theory cannot properly explain this paradox other than saying the astronaut will age more slowly! For this reason, I think we need a more reliable method for measuring time.

I think what's causing confusion for everybody, including Einstein, is that the regular motion of objects and the motion of light are different things. Regular motion involving objects and moving persons are subject to what is called perspective or simply geometry. Ask any artist or a person who does technical drawing and they will tell you that one of the very first things they learn is perspective.

Consider a window with 2 half open wings (illustrated in the Figure below). When you stand between the wings, they will appear wide open in your reference frame. If you move parallel to the wall, you will see that the wing on the path of your movement appears to be closing and the other one opening even though there is no physical force acting on the wings to cause them to move. This effect arises from the point of the moving observer's perspective: the window's wings move in space in the opposite direction of the observer's movement, but they will still be half open in reference to the wall.

Figure 3.5   A window with 2 half open wings

In the Time Dilation theory definition, which is based on Einstein's special theory of relativity, a stationary clock consisting of a light pulse (emitted vertically into the space) bouncing between two mirrors, the light assumed to be moving diagonally in the opposite direction for an observer moving relative to the clock, therefore ticking at a slower rate for the moving observer. From the view of geometric perspective, this effect is similar to the window example I gave. But, we cannot ignore the fact that the observer is moving. If he was standing still and the light was moving diagonally, then the clock would tick slower and I wouldn't argue with that. Because it would take more time for the light to propagate the diagonal distance which is obviously longer than the vertical distance. But, since the observer is moving, it can be said that each point in the path of the light is moving at the same speed away from the observer. The light still moves vertically (independent of the observer's movement as postulated by Einstein himself).

Let's do a simple calculation. If the observer is moving to the left with a speed of 1 km/s, the light reflected from the bottom mirror A will move vertically with a constant speed of 300.000 km/s. Both mirrors, bottom mirror A and top mirror B will move in the opposite direction from the observer with a speed of 1 km/s. Thus, if the distance L between the mirrors is 300.000 Km, the light moving vertically will hit the center of the top mirror B exactly after 1 second. This 1-second interval is same both for the moving observer and another observer standing by the stationary clock. Therefore, time dilation won't occur.

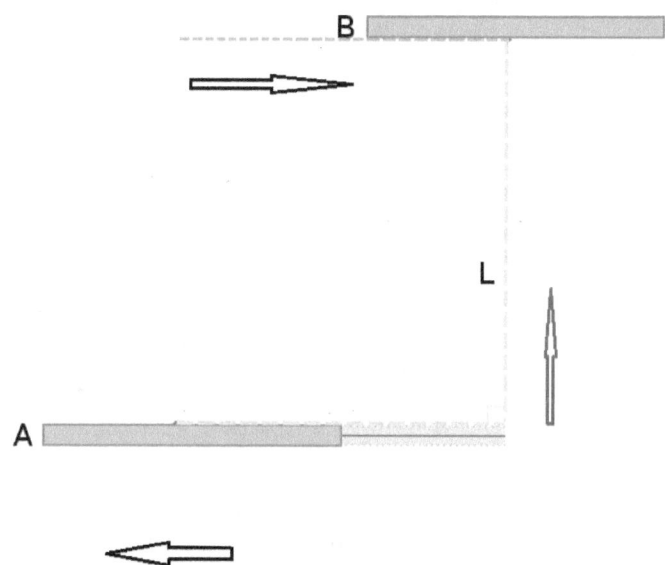

Figure 3.6 A stationary clock device consisting of vertically emitted light beam bouncing between two parallel mirrors as observed by a moving person

To double check and confirm my reasoning, let's consider this time a clock consisting of a mechanical piston moving vertically up and down with a speed of 1 meter/sec.

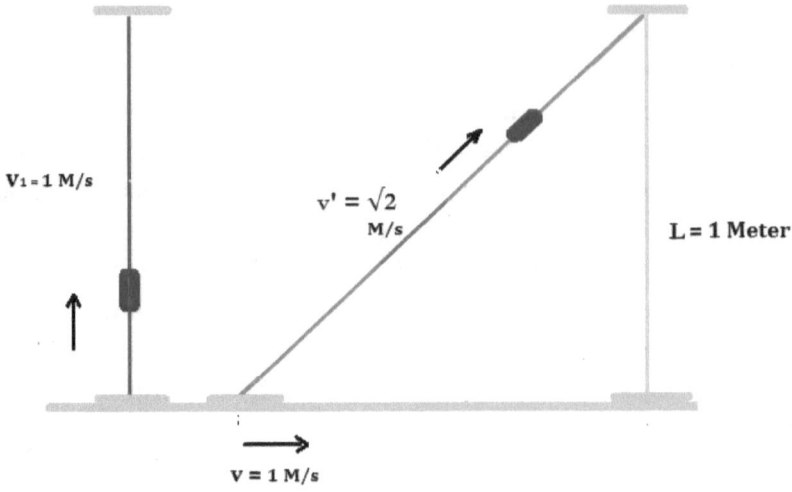

Figure 3.7 Mechanical piston clock device

If the observer is moving relative to this mechanical clock with a speed of 1 meter/s, the piston will also have a horizontal speed in the opposite direction from the observer. The piston will move diagonally in the opposite direction from the observer with a speed of v' = √2 m/s. This is the expected result achieved by applying the Newton's law on projectile motion and geometry, confirming that the clock won't slow down for an observer moving relative to a stationary clock because the piston will move the diagonal distance of √2 meters in exactly 1 second (gravity (g) acting on the piston is negligible, because the piston being part of a mechanical clock has a constant vertical speed of 1 meter/s).

It is very clear from the two examples I gave above that in order for the vertically moving light to have a diagonal direction, according to the Newton's law of motion and the Pythagorean theorem in geometry, a horizontal velocity should be added to the speed of the light. If this could be possible, the light would have a diagonal velocity faster than 300.000 km/sec - the universally accepted maximum speed the light can move - which I am sure everybody will agree with me that it is not possible. Therefore, there is no point of further arguing that time would slow down from the perspective of a moving person in reference to a stationary clock. It certainly doesn't slow down in the mechanical clock example I gave above. Time is universal and its rate is constant for all observers moving relative to each other in space.

Allow me to go back to the time dilation theory to explain why a light-clock device should not be used in the first place. If the clock device is carried by the observer, the light emitted by the clock would still move vertically, because the light wouldn't be affected by the speed of the observer. This simply means that both clocks, clock carried by the observer and the stationary clock would tick at the same rate and there won't be time dilation. It's logical to conclude that a clock consisting of a light bouncing between two mirrors cannot really be moved. After some time the clock would cease to tick because the light bouncing away from the mirrors carried by the moving observer, at one point would keep moving into the open space; the top mirror won't be on its path to reflect it back.

My examples show that time cannot be the 4th dimension of the space-time continuum because the speed of light (a temporal event) is independent of any inertial frame, so is time. Therefore, Einstein's Special Relativity and Spacetime theories are fundamentally flawed. Einstein had based his theories on the assumed foundation that the observed rate at which time passes for an object depends on the object's velocity relative to the observer. My logical reasoning based on my examples proves otherwise. Time is not the 4th dimension of the space-time continuum, as it is independent of events that occur in the universe. Time can be used as a master tool or reference to measure, record and classify temporal events, but is not a physical dimension. Hypothetically, each person can create his/her own time reference frame, by starting from 'Now' and using any length of interval he/she chooses. But, as humans, we constantly interact with each other, therefore for practical reasons we adapt to a world-wide accepted time standard. I have included beloew the mathematical formula that proves my logical reasoning for those of you who may want to look at it.

Mathematical proof: No physics theory would be complete without a mathematical formula. Let's study the formula used to calculate time dilation:

$$\Delta t' = \Delta t / \sqrt{(1-v^2/c^2)}$$

Without going into a deep analysis, from the first glimpse it's very clear that this equation is fundamentally flawed. Because, when you have the v^2/c^2 ratio in your formula you are making the speed of light (c) depend on the speed of the person who is moving (v). Remember that the speed of light c with respect to any inertial frame is a constant and is independent of the motion of the light source. I have already demonstrated in the main article that c (speed of light) cannot be used for time measuring purposes from the reference point of a moving person, **unless** it's proven that the speed of light c with respect to any inertial frame is not constant and is not independent of the motion of the light source. Therefore, the time dilation formula is invalid.

My own formula (based on the mechanical clock example I gave above)

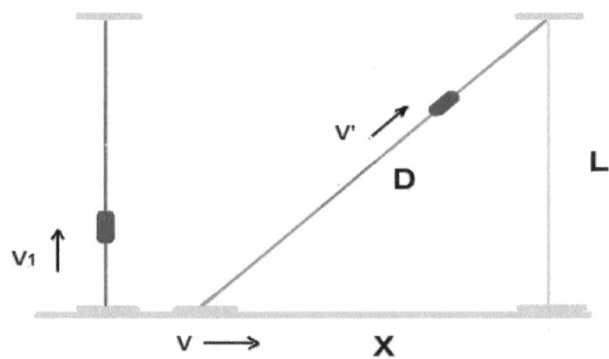

$D = \Delta t'. V'$ $X = \Delta t'. V$ $L = \Delta t. V_1$ (D, X, L are the sides of the triangle)

$V'^2 = V^2 + V_1^2$ (Newton's law on projectile motion)

$D^2 = X^2 + L^2$ (Pythagorean theorem)

$(\Delta t'. V')^2 = (\Delta t'. V)^2 + (\Delta t . V_1)^2$

$\Delta t'^2 . V'^2 = \Delta t'^2 . V^2 + \Delta t^2 . V_1^2$ (replacing V'^2 with $V^2 + V_1^2$)

$\Delta t'^2 . (V^2 + V_1^2) = \Delta t'^2 . V^2 + \Delta t^2 . V_1^2$

$\Delta t'^2 . V^2 + \Delta t'^2 . V_1^2 = \Delta t'^2 . V^2 + \Delta t^2 . V_1^2$

$\Delta t'^2 . V_1^2 = \Delta t^2 . V_1^2$

$\Delta t'^2 = \Delta t^2$

$\Delta t' = \Delta t$

Proving that the passage of time is equally same for an observer (moving in reference to a stationary clock) and another person (standing by the clock). Thus, my formula proves that time doesn't slow down from the perspective of a person who is moving relative to a stationary clock.

CONCLUSION

My mathematical formula proves that the passage of time is equally same for an observer (moving in reference to a stationary clock) and another person (standing by the clock). Thus, my formula proves that time doesn't slow down from the perspective of a person who is moving relative to a stationary clock. As far as I am concerned, Einstein's Special Relativity and Spacetime theories are fundamentally flawed. Einstein had based his theories on the assumed foundation that the observed rate at which time passes for an object depends on the object's velocity relative to the observer. My mathematical formula based on my example proved otherwise. Special relativity postulated that the time dilation effect is reciprocal: as observed from the point of view of either of two clocks which are in motion with respect to each other, it will be the other clock that is time dilated. This statement contradicts itself and therefore it's a false statement; which means that time dilation is an observed effect and it doesn't actually occur in reality.

On the other hand, Einstein's theory of general relativity postulated that the strength of gravitational fields can slow the passage of time for an object as seen by an observer outside the field. But, we have to consider the fact that a gravitational field is a type of force. It is possible to alter the rate of a clock by applying a force to it. I don't argue with that, but this is not time dilation in relativistic contexts, it is mechanical malfunctioning. To prove my argument, I take the statement postulated by the theory that the gravitational time dilation is not reciprocal: an observer at the top of a tower will observe that clocks at ground level tick slower, and observers on the ground will agree about the direction and the magnitude of the difference. The theory further postulated that there is still some disagreement in a sense, because all the observers believe their own local clocks are correct, but the direction and ratio of gravitational time dilation is agreed by all observers, independent of their altitude. The way I see it, this disagreement arises from the fact that gravitational time dilation is just mechanical malfunctioning as I have argued above. The mathematical formula used by relativists for the gravitational time dilation was based on the special relativity formula which has the v^2/c^2 ratio. I have already argued above and declared that the formula is invalid. Therefore, the general relativity is also mathematically an invalid theory.

Time is not the 4th dimension of space-time continuum as it is independent of events that occur in the universe. Time can be used as a master tool or reference to measure, record and classify temporal events, but is not a physical dimension. Hypothetically, each person can create his/her own time reference frame, by starting from 'Now' and using any length of interval he/she chooses. But, as humans, we constantly interact with each other, therefore for practical reasons we adapt to a universally accepted time standard.

ABOUT THE AUTHOR

Billy Coskun is a scientific philosopher, inventor, software developer, digital artist and electronic music composer. In 2015, he has developed the Electrum Quantum Audio Engine app which is the world's first commercial quantum product. Billy Coskun's experiments with the quantum audio engine also supported his findings regarding the super-deterministic virtual multiverse. Therefore, the Quantum Audio Engine app is recommended to the readers to confirm Billy Coskun's findings - available on Google Play and Samsung Galaxy app stores.

ATTRIBUTION

Figure 1.1　Gnixon - Public Domain

Figure 1.2　S Sepp - Public Domain

Figure 2.1　Anonymous work

Figure 2.2　LucasVB - Public Domain

Figure 2.3　Public domain

Figure 3.1　Public Domain

Figure 3.2　Acdx - Public Domain

Figure 3.3　Acdx - Public Domain

Figure 3.4　Binarysequence - Public Domain

Figure 3.5　Billy Coskun - Own work

Figure 3.6　Billy Coskun - Own work

Figure 3.7　Billy Coskun - Own work

www.ingramcontent.com/pod-product-compliance
Lightning Source LLC
Chambersburg PA
CBHW021019180526
45163CB00005B/2024